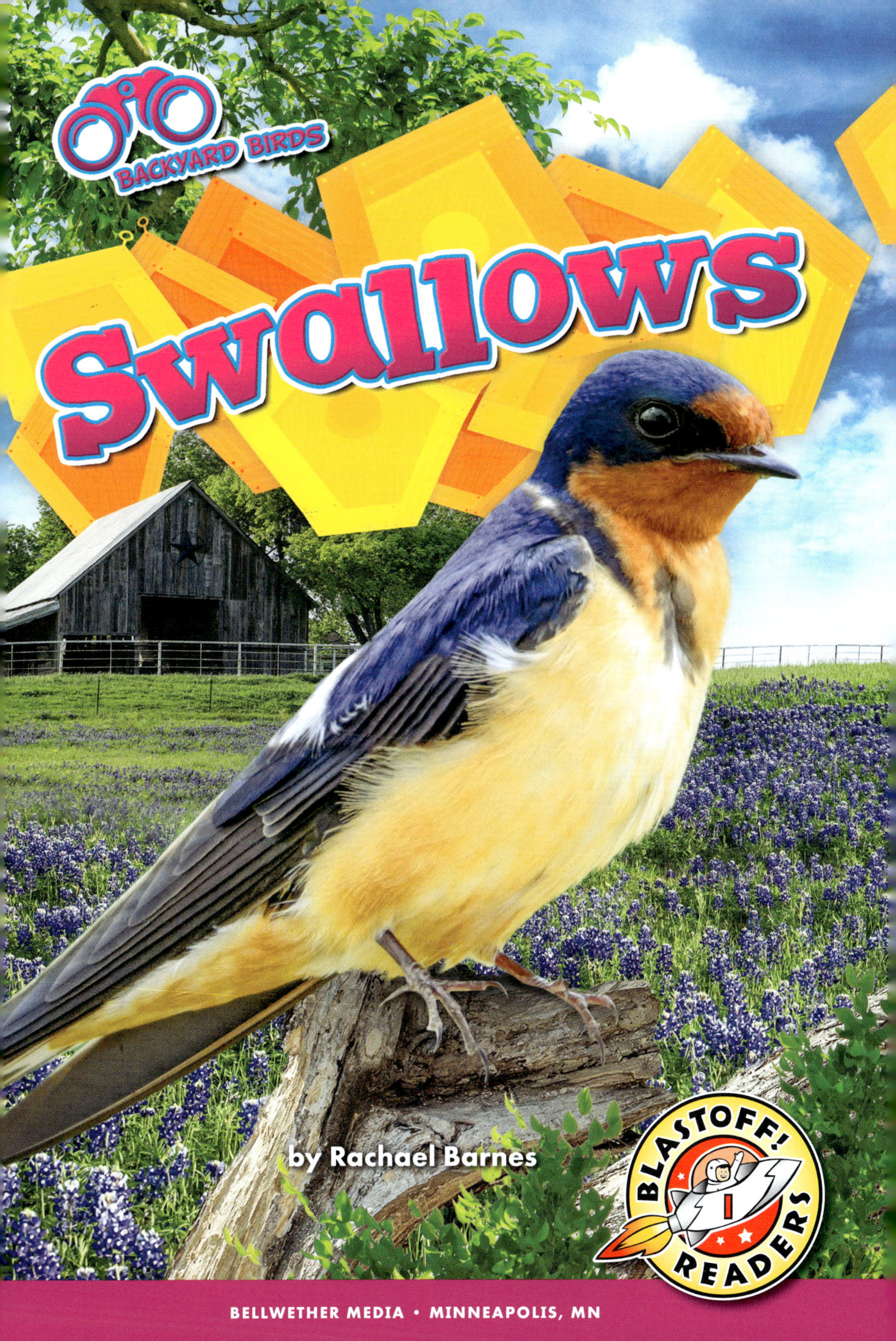
BACKYARD BIRDS
Swallows
by Rachael Barnes
BLASTOFF! READERS
1
BELLWETHER MEDIA • MINNEAPOLIS, MN

Blastoff! Readers are carefully developed by literacy experts to build reading stamina and move students toward fluency by combining standards-based content with developmentally appropriate text.

Level 1 provides the most support through repetition of high-frequency words, light text, predictable sentence patterns, and strong visual support.

Level 2 offers early readers a bit more challenge through varied sentences, increased text load, and text-supportive special features.

Level 3 advances early-fluent readers toward fluency through increased text load, less reliance on photos, advancing concepts, longer sentences, and more complex special features.

★ **Blastoff! Universe**

Reading Level

This edition first published in 2023 by Bellwether Media, Inc.

Library of Congress Cataloging-in-Publication Data

Names: Barnes, Rachael, author.
Title: Swallows / Rachael Barnes.
Description: Minneapolis, MN : Bellwether Media, 2023. | Series: Backyard birds | Includes bibliographical references and index. | Audience: Ages 5-8 | Audience: Grades K-1 | Summary: "Developed by literacy experts for students in kindergarten through grade three, this book introduces swallows to young readers through leveled text and related photos"– Provided by publisher.
Identifiers: LCCN 2022002376 (print) | LCCN 2022002377 (ebook) | ISBN 9781644876930 (library binding) | ISBN 9781648347399 (ebook)
Subjects: LCSH: Swallows–Juvenile literature.
Classification: LCC QL696.P247 B37 2023 (print) | LCC QL696.P247 (ebook) | DDC 598.8/26–dc23/eng/20220125
LC record available at https://lccn.loc.gov/2022002376
LC ebook record available at https://lccn.loc.gov/2022002377

Editor: Rebecca Sabelko Designer: Laura Sowers

Printed in the United States of America, North Mankato, MN.

Table of Contents

What Are Swallows? 4

Swooping for Insects 8

Life in the Sky 12

Glossary 22

To Learn More 23

Index 24

What Are Swallows?

Swallows are small **songbirds**. They are in the swallow family.

All in the Family
cliff swallow
purple martin
tree swallow

Swallows often have blue or green feathers. Most have **forked tails**.

forked tail

Swooping for Insects

Swallows **swoop** over lakes and fields. They like to live near water.

Swallows look for food as they fly. They eat a lot of **insects**.

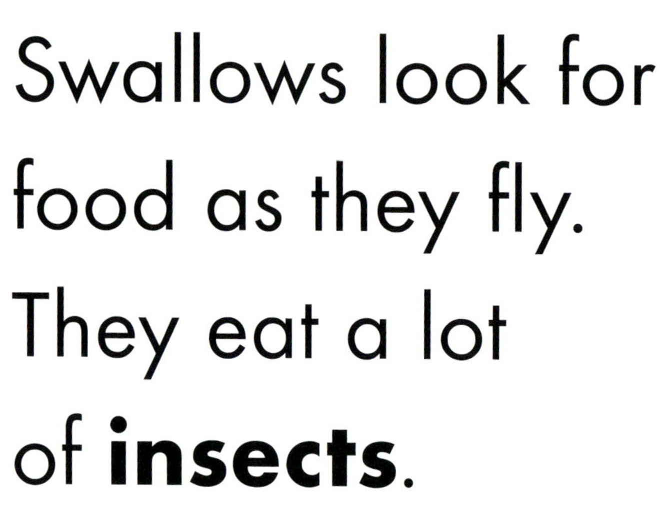

Swallow Food
insects
berries

Life in the Sky

Many swallows make mud nests on buildings. Others make nests inside trees.

tree nest
mud nest

Swallows chatter often. Some calls tell other birds about danger.

Swallow Call

cheep!

churee!

Some swallows live in **flocks**. They fly together in huge circles.

flock

Swallows **migrate** for warmth and food.

These birds live much of their lives in the air!

Glossary

flocks

groups of birds

migrate

to travel with the seasons

forked tails

tails with two parts that are shaped like the letter Y

songbirds

birds that make musical sounds

insects

small animals with six legs and hard outer bodies

swoop

to quickly fly down

To Learn More

AT THE LIBRARY

Addison, Amanda. *Boundless Sky.* Oxford, U.K.: Lantana Publishing, 2020.

Barnes, Rachael. *Crows.* Minneapolis, Minn.: Bellwether Media, 2023.

Neuenfeldt, Elizabeth. *Chickadees.* Minneapolis, Minn.: Bellwether Media, 2022.

ON THE WEB

FACTSURFER

Factsurfer.com gives you a safe, fun way to find more information.

1. Go to www.factsurfer.com.
2. Enter "swallows" into the search box and click 🔍.
3. Select your book cover to see a list of related content.

Index

buildings, 12
calls, 14, 15
chatter, 14
colors, 6
family, 4, 5
feathers, 6
fields, 8
flocks, 16, 17
fly, 10, 16
food, 10, 11, 18
forked tails, 6, 7
insects, 10
lakes, 8
migrate, 18
nests, 12, 13
size, 4
songbirds, 4
swoop, 8
trees, 12, 13
water, 8

The images in this book are reproduced through the courtesy of: Jeff W. Jarrett, front cover (swallow); NicholasGeraldinePhotos, front cover (backyard); WildlifeWorld, pp. 3, 22 (songbirds); Tom Reichner, pp. 4-5; Sharon Haeger, p. 5 (cliff swallows); Agami Photo Agency, p. 5 (purple martin); Paul Roedding, p. 5 (tree swallow); All Canada Photos/ Alamy, pp. 6-7; Werner Baumgarten, p. 7 (forked tail); PatP66, pp. 8-9; Fiona M. Donnelly, pp. 10-11; Luc Pouliot, p. 11 (insects); Hinochika, p. 11 (berries); Pascale Gueret, pp. 12-13; Hayley Crews, p. 13 (tree nest); GiPilr, pp. 14-15; William Leaman/ Alamy, pp. 16-17; imageBroker/ Alamy, pp. 18-19; Jiri Fejkl, pp. 20-21, 22 (swoop); rck_953, p. 22 (flocks); Drakuliren, p. 22 (forked tails); Winai Pantho, p. 22 (insects); Nicolae Cirmu, p. 22 (migrate); Gallinago_media, p. 23.